How To Build an
ECO-FRIENDLY GREENHOUSE

Reduce Reuse Recycle

ANDROS ZACHARIA

AuthorHouse™
1663 Liberty Drive
Bloomington, IN 47403
www.authorhouse.com
Phone: 1 (800) 839-8640

© 2015 Andros Zacharia. All rights reserved.

No part of this book may be reproduced, stored in a retrieval system, or transmitted by any means without the written permission of the author.

Published by AuthorHouse 06/10/2015

ISBN: 978-1-5049-0136-9 (sc)
ISBN: 978-1-5049-0137-6 (e)

Library of Congress Control Number: 2015906014

Print information available on the last page.

Any people depicted in stock imagery provided by Thinkstock are models, and such images are being used for illustrative purposes only.
Certain stock imagery © Thinkstock.

This book is printed on acid-free paper.

Because of the dynamic nature of the Internet, any web addresses or links contained in this book may have changed since publication and may no longer be valid. The views expressed in this work are solely those of the author and do not necessarily reflect the views of the publisher, and the publisher hereby disclaims any responsibility for them.

Table of Contents

Acknowledgment 1

Introduction 3

Chapter 1   Benefits of an Eco-Friendly Greenhouse 5

Chapter 2   Choosing, Collecting, and Cleaning Bottles 6

Chapter 3   Materials 8

Chapter 4   Tools 10

Chapter 5   Costs Associated with the Greenhouse 11

Chapter 6   The Construction 12

Chapter 7   Tips and Recommendations 32

Chapter 8   A Peek inside our Greenhouse 38

Chapter 9   The Green Initiative 40

Appendix A Detailed Materials List 44

Appendix B Preparing the base area 45

About the Author 51

Acknowledgment

Before I start thanking those individuals or organizations who actually assisted me in putting this together, I would like to thank those persons who came before us. Their courage and giant leaps into the unknown have paved the way for us to be creative. So I thank all those people who I have not met or known who have made this possible. I also thank life for whatever it has given me and not given me.

First and foremost I would like to express my gratitude to my father Anastasis Zacharia without whom this book would not have been written. My father was the individual who built our greenhouse with my support which, without his knowledge and expertise as a handy man, would not have been possible. However, without the support of my mother Maria Zacharia, neither of us would have been able to build the greenhouse or write this book as she continuously fueled us with love and understanding which are part of the cement of this manual. My parents have served as the backbone in my life, and it's because of them this has been possible to achieve.

With capital letters I would like to thank the following persons for their support and believing in me:
David Lee Andrews
Fabien Amine
Yiota Mallas

I would like to thank the following individuals for their contribution but also sticking to the same brand of water:
George Christofi
Howard and Pat Jordan
Marios Souroullas
Sandra Benko

I would like to thank the following companies::

GreenDot Cyprus, for allowing us to get dirty and collect plastic bottles from their refuse collection yard.
Saint Nicholas Water Company for sharing information with us and providing us with used bottles.
Tremythos Restaurant in Ayia Anna

Introduction

The Republic of Cyprus is an island that consumes a lot of plastic-bottled water due to the exceptionally warm climate, but also because of the widespread assumption that tap water is not safe to drink. Plastic bottles are made of polyethylene terephthalate (PET), which is the primary material from which food and beverage containers are made. There are many companies in Cyprus that produce bottled water, and there are also a lot of water brands that are imported from other countries. In general, many companies prefer plastic as the best alternative to glass due to its inherent light weight, low cost of production and transportation, and resistance to breakage.

To my surprise, when first researching the impact of plastic bottles in Cyprus, I originally thought that plastic water bottles were recycled by the water companies, but due to health requirements, the actual water bottles are made from virgin plastic.

According to Saint Nicholas, a Cypriot water bottle company, it is estimated that 100 million plastic bottles of all types and sizes are produced in Cyprus per year. Based on data collected by the Republic of Cyprus Department of Statistics, the total number of plastic units imported into Cyprus (which includes carboys, flasks, bottles, and other similar articles of plastic) was 81,274,370 units (1,804,167 kg) in 2010. This number fell to 65,146,059 units in 2013.

Initially, before we considered putting together our eco-project, our plan was to assemble a garden greenhouse using new materials. The idea to use plastic bottles arose when I started my own blog which focused on eco-concepts and well-being. After writing about various products and ideas through my site, we then considered using plastic bottles to make the greenhouse. We knew there was an abundance of bottles available, so our plans were just a step away from materializing.

Initially we knew that this was going to take some time to put together, but the advantages of it would definitely outweigh the time invested and the effort required to make this happen. This project would be a small contribution towards helping what was already being done in preserving and sustaining our environment. Although Cyprus has implemented various recycling systems and other green projects, incorporating a green culture is still in the early stages for this country. It's one thing to be told what to do, and it's another to act on your own initiative. Once people learn to operate consciously on a green level, that is when it ceases to be *part of* their culture, because it *is* their culture.

Apart from doing our bit for mother earth, we knew that during cold periods, we would be in a position to provide for ourselves and to support a healthy lifestyle.

The purpose of this guide is to explain how through this application process, you may create any structure you want, not just a greenhouse. We created a greenhouse because it was something we required.

Chapter 1
Benefits of an Eco-Friendly Greenhouse

As with any project, appropriate planning, time, and effort are required to ensure its effective completion. This is a challenge, however, the many benefits exceed the costs:

- Creating something is very rewarding in itself because what you have created speaks for itself. Before starting, the preparation phase seemed very lengthy, but we eventually did it.

- Money does not grow on trees; money is earned through hard work. The choice was either to pay for a glass greenhouse or to put some labor into it. Our costs were next to nothing, by contrast with having a proper greenhouse installed, which can be quite expensive. What we would have spent has now been saved.

- Because we utilized used plastic bottles to make it, we knew we were being more environmentally responsible, without contributing further to the carbon footprint. We all have a duty to the environment, especially since it's where we live. The more we look after the environment, the more it will look after us. It is common knowledge that plastic is a harmful material, and due to its unnatural composition, it is not easily biodegradable like most materials.

- By utilizing used plastic bottles, we saved on landfill space, even if only by a slight percentage. In addition to bottles taking up space, they take over three hundred years to biodegrade.

- We are capable now of providing for ourselves 100 percent of the year. We enjoy unseasonal vegetables throughout the year, especially during the winter period. What we have in the summer, we now have in the winter, which is the benefit any greenhouse provides. Plastic is a good form of insulation—better than many other materials.

- Having tried a project of this nature, you will become more capable and confident to take on other projects.

Chapter 2
Choosing, Collecting, and Cleaning Bottles

Out of the whole process, finding and collecting the bottles was the most enjoyable. Even though it was fun, it was not easy collecting them. One of the challenges we faced was finding the same bottles, because people buy various brands, and not all brands are the same; in fact, no one brand had the same bottle shape. We knew what special offers were on for a particular month or week as brand choice was linked to cost.

Choosing the Bottles

It is important to decide which bottles to use for your structure. The reason being that some bottles serve as a better form of insulation than others. When fit together, square bottles sit perfectly, making the structure more windproof. We found that the narrower necks of some round bottles created gaps when aligned next to each bottle strip. Because of these factors, we used Saint Nicholas water bottles that varied from 0.5 liters to 1.5 liters.

Depending on your time constraints and which bottles are available for use, you may be forced to use certain bottles you were not planning to. Make do with what you have available, and then you can replace the bottles once you have enough of the single type you prefer or that works better. We do recommend ways to make your greenhouse as windproof and insulated as possible.

Collecting the Bottles

After we decided which bottles we were going to use, we had to plan how to get them. This is how we collected the 2,500 bottles for the project.

- **Facebook**—Using Facebook, I was able to reach a larger audience by asking friends to save their water bottles for me. There are always those few willing to help you. Try it; it doesn't cost a penny.

- **Recycling yards**—We visited the refuse recycling yard of a company called Green Dot to collect bottles. This proved to be a plastic gold mine for us. The process was not very clean, so be prepared to get dirty.

- **Local restaurants and bars**—Another good source for collecting bottles is bars and restaurants. What really helps is to find out which brand of bottles certain places sell. I was able to identify specific places by asking at the actual bottle companies that supply these places. Some places were willing to work with us; others, not at all.

Cleaning the Bottles

You will find that some bottles may need a little rinse with some lukewarm water to clean them. Part of the preparation of the bottles requires you to cut off the paper label around them and remove the bottle caps. Do not throw away the bottle caps as these can be used for something else or by someone else. This process can get quite messy if you are not organized, so make sure you have designated bags in which to place each item.

Chapter 3

Materials

The materials you utilize may vary from project to project and person to person. This is absolutely up to you. Just for your own information, we used the following materials for this project:

Timber—Treated timber was used for the structure of the greenhouse. Make sure your timber is straight. We used 2 inch x 2 inch planks. The timber was treated to protect against weather conditions as it may decompose if not protected.

Bamboo sticks—These are used to make the actual bottle strips, keeping the bottles aligned and straight. They were obtained locally from the countryside. It is important that the bamboo sticks are as straight as possible. Not having straight bamboo sticks will create gaps between bottles when they are placed side by side. The choice of using the bamboo stick was due to the fact that when they are cut they re-grow at a very fast rate.

Saddle clips—These are used to hold down the bamboo sticks. They can be purchased from any Do-It-Yourself (DIY) store. Make sure that the brackets fit the bamboo stick. Check the size of your bamboo sticks and make sure they are not loose once the clips have been screwed over them.

Plastic bottles—The type of bottle(s) you decide to use is totally up to you. This depends on what you have access to and how many you are able to get. Some bottles are better than others, but you will have the final say.

Screws—Two different types of screws are required. You will need screws to fix the saddle clips, as well as larger screws that are long enough to join the different sections of the timber frames together during the assembly phase, but also when making each section.

Nylon sheets—These are necessary to keep the greenhouse more insulated, but they are also used on the roofs to prevent water passing through. You can see this process later on in this book.

See appendix A for quantities of the above.

Chapter 4
Tools

Having access to the right tools can make your work easier, however, we do not want you to spend too much money on tools when assembling your greenhouse.

The tools that we used to create and assemble the eco-greenhouse were:

- battery-powered drill, including small general-use drill bit and plastic/wood hole drill bit

- carving knife

- screwdriver (Phillips or straight, depending on the screws used on the saddle clips)

- tape measure and pencil

- saws (small and big)

You could try asking friends or neighbors if they have any of the tools that you can use. Make sure you return them in the same condition they were in when you received them. You may even try buying secondhand tools by looking for them online or by visiting local secondhand stores in your area. Try to be as economical as possible.

When using the tools, you need to follow health and safety requirements to avoid any accidents. Make sure you wear the necessary eye-protection glasses and gloves where required.

Chapter 5
Costs Associated with the Greenhouse

Some costs are incurred during this process, whether it involves taking a drive to pick up some bottles from someone's house or using up electricity to charge the battery pack for your drill. What's important is to minimize it as much as possible. Arranging same-day bottle pickups will help you save on fuel as well as reducing your carbon footprint.

You might consider asking certain local companies if they would be willing to donate some materials in support of an eco-friendly project. You could also ask a company to sponsor you by paying for the expenses, and in return, you could advertise the company name when you write about your project or if it is mentioned in a local newspaper. We were fortunate because we had these tools ourselves.

Make it in the most inexpensive way possible.

Chapter 6
The Construction

We have tried to give a descriptive account of the process of making the greenhouse from beginning to end. Once you have read through this and studied our illustrative pictures, you will have an overview on how to proceed with your very own plastic bottle greenhouse—or any type of plastic structure, for that matter. Please bear in mind that each person may tailor a greenhouse according to his or her requirements and preferences for the size and components like windows, doors and roof size. Do what is convenient and will be useful for you.

All required materials should be purchased once you have finalized the specifications of your greenhouse, ensuring that you buy exactly what is required (otherwise you may buy too much or too little). Remember, you want to try to be as cost-effective as possible.

Sketching Your Design

Like any structure, some basic design plans need to be put on paper. A simple sketch of your structure with likely measurements will give you a rough guide of how many bottle strips you will need for each part of the greenhouse. In total, our greenhouse has eight separate parts including the main beam for the roof: 2x main wall sections, 2x side wall sections, 2x roof sections, 2x side roof sections and 1 roof beam. Whether you will have any windows is something to consider at this stage. However you can easily install a window at a later point if you feel it is necessary.

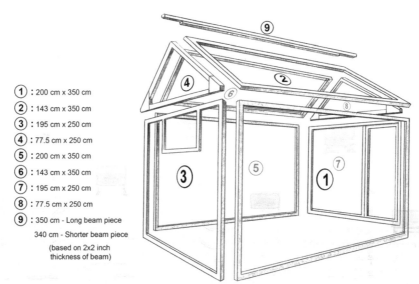

① : 200 cm x 350 cm
② : 143 cm x 350 cm
③ : 195 cm x 250 cm
④ : 77.5 cm x 250 cm
⑤ : 200 cm x 350 cm
⑥ : 143 cm x 350 cm
⑦ : 195 cm x 250 cm
⑧ : 77.5 cm x 250 cm
⑨ : 350 cm - Long beam piece
340 cm - Shorter beam piece
(based on 2x2 inch thickness of beam)

- **Working Area**

Have a designated working area. You will need a chair, especially for when you will be cutting and putting together the bottles. Make sure you have a bucket or plastic rubbish bag so you can put the various cuttings in. Keep it clean.

- **Frame Work**

We recommend that you put your frames together prior to making your bottle strips. Again, the frame designs will depend on whether you will have a window or where you would like to put the door and/or window. As an example, you can see from our greenhouse, we placed the door and window opposite each other. The reason for this was to create ventilation when both opened. How you decide to proceed is up to you. After each frame was prepared, we installed the bottle strips. The images below show some frames.

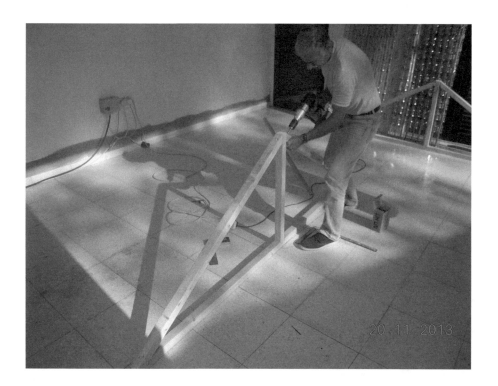

Making and Installing Bottle Strips

The most consuming part of the process is making the bottle strips. By this point, hopefully you will have collected the majority of the bottles required for your greenhouse, but if you haven't, do not be disheartened. At least have a sufficient number so you can start your project. In the beginning, this process can be quite slow; eventually, though, you will become swift. Following is a step-by-step process of how to make the bottle strips:

Step 1: Cutting bottles
Take a bottle and cut carefully 5.5 cm from the bottom as evenly as possible. Do the same for each bottle. Some bottles are used at the end of the strips and will not need to be cut (see Step 3).

Step 2: Joining bottles
Once you have cut enough bottles to make up the height of the frames, the bottles will be joined to make the strips. You will need to place your hand in the bottle so that you can fit one bottle into the other easily. Be careful as you do not want to damage the bottle in any way. The bottles can be adjusted to ensure that you have the right length. See the following illustration of this process:

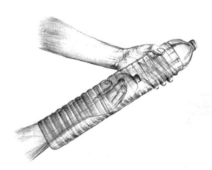

Step 3: Rear-end bottle preparation and fitting

The end bottles on each strip will not have their bottoms cut off. This allows them to rest comfortably on the frame. Before you cut off the necks of the bottles (as illustrated below), puncture a hole in the center of the bottom of each bottle that is just big enough to insert the bamboo sticks (or other material, if you will use an alternative). The two illustrations below show the puncturing of the rear end of the bottle. The first illustration on the left shows the drilling of the hole using a small drill-bit. This will make it easier and safer to then use the bigger plastic hole drill-bit to enlarge the hole. Some people may prefer to use the plastic hole drill-bit directly instead of using the small drill-bit first.

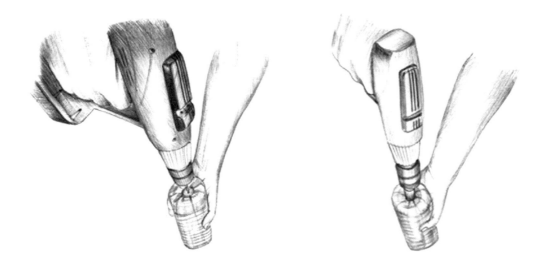

The following illustration shows the type of drill-bit used as well as the hole it has punctured in the bottom of the bottle. When drilling a hole, you must make sure it is big enough to insert a bamboo stick or equivalent. The stick should fit snugly through the hole.

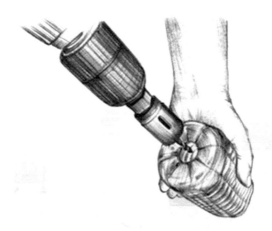

Once the holes have been made, the necks of the bottles must be removed. The illustration shows roughly how much neck you should cut off. Do not cut too much. It is better to cut less and then adjust accordingly. By cutting too much you will be unable to fit the bottle securely into the other bottle whilst having enough length to rest on the frame.

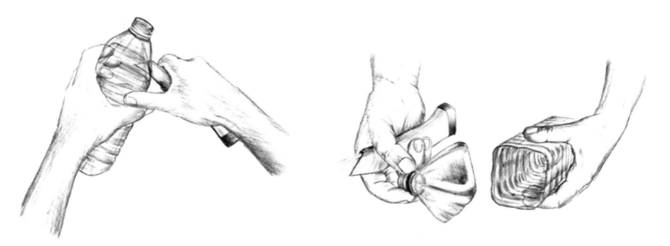

Once the two rear-end bottles have been punctured and necks removed, these must be fitted to the other bottles. How you decide to fit one into the other is up to you. Below is an illustration of how the rear bottle has been fitted. Notice that we take the bottle from its rear and slowly insert it over the neck of the next bottle.

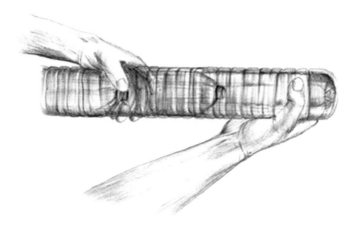

Once all bottles have been fit together, you can now place the bamboo stick through the strip of bottles. Bamboo sticks need to be as straight as possible. If you are using bamboo sticks with foliage, make sure the foliage has been removed before placing it through the bottles.

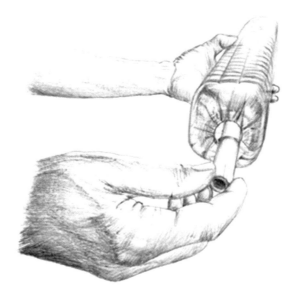

Step 4: Installation of bottle strips
Once you have made your strips, they can be installed. To avoid gaps, each strip is placed on the frame so it is tightly packed next to other bottle strips. Take a saddle clip and place it over the end of the bamboo stick and screw it down to the frame. See the image below.

The excess parts of the bamboo sticks can be cut off once you have installed them, *but not before*.

This process becomes a little challenging when fitting the bottles to the side roof sections and window frame. Remember, the bottle strips are made in the same manner for any section, despite the shape of the frames (i.e., triangle). Again, the design of the greenhouse may vary, so your needs will change. Sometimes a little innovative thinking is needed along the way. The illustration below shows how we modified our bottles to fit within the frame. We used some small 0.5 liter plastic bottles.

You may notice some gaps at the corners of some of the frames, especially for the triangular sections, because it is difficult to place additional bottles due to limited space. We simply used flat woodchip material, just enough to cover the opening. See the image below.

Sections ready for assembling: once you have installed the bottle strips and your cross-sections are all covered, you can assemble the greenhouse. When your sections are complete, you should have something that looks similar to the following images. The first image is the side section with door. The next image shows the triangular sections forming the roof.

The next image shows what the walls of the greenhouse as well as the roof sections will look like with the nylon cover. The image following that shows the second side wall with a window.

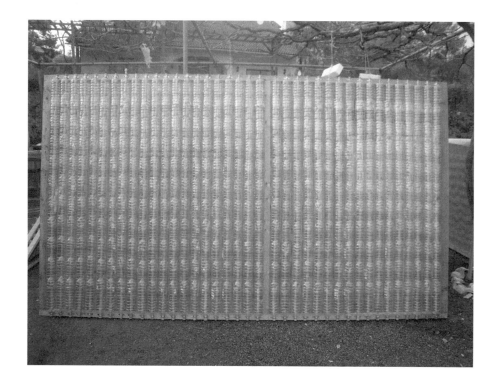

Step 5: Making the base section

Some preparation needs to be made to the area where the greenhouse will be assembled. It is important that it sits on a surface other than the actual ground itself, because you will find that over time, timber will start to decompose, especially if it is in direct contact with the soil. This part of the process can be adapted according to each person's preference.

In order to have good base on which the greenhouse will sit, you need to make some accurate measurements. We used a simple, inexpensive, accurate technique referred to as the water-level method. Please see Appendix B for further details on this method. You are not limited to this method, use any method that will give you the same result.

Step 6: Assembling the greenhouse

Once the base area has been prepared, you can begin to put together the different sections. During this phase of the process, you will need at least three people. It took us roughly four hours to assemble the greenhouse.

First you will need to attach the rear wall with one of the side walls to create a right angle. One person should hold the one side while another person holds the other section. A third person then screws together the sections, which will become an L-shaped section that can stand alone. See Illustration 2.

Once the first two sections are up, the second side wall may then be attached to the unit. See the following illustration.

As soon as the two side sections and back-wall are up, then the two triangulated side sections can be attached to the structure. You will notice that a metal bracket has been used at the bottom of the triangulated section. This is to secure the bottom section and top section together. This has further been secured at the two ends using screws.

The greenhouse is now ready for the placement of the two roof sections. However, before proceeding, we will need to place a beam from one side to the other to support the roof sections. The beam consists of two pieces of timber and a bracket in the middle upon which the roof sections will rest.

The illustration below gives a clearer presentation of the beam.

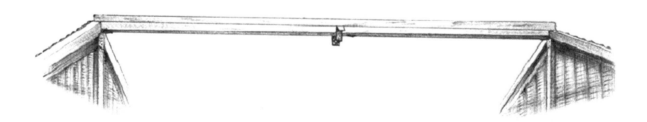

Just two more sections to go, and we have completed the greenhouse. At this point, remove any objects that are inside the greenhouse before attaching the last two sections. Then attach the other roof section before putting on the main outside wall.

The image below shows the screw drilled into the frames, securing the sections.

Chapter 7
Tips and Recommendations

Throughout the process of building the plastic bottle greenhouse, we managed to generate ideas to optimize our greenhouse to the best of our ability to achieve an effective result. The following points may be useful as you work on your project, however, you may apply or adapt what you believe is best for you.

Tip 1: Larger bottles

Using larger bottles (for example, two-liter size) would be the most effective for a greenhouse. Larger bottles work as a better form of insulation and help to maintain higher temperatures. Using a larger bottle, you will cover your surface sections much quicker, and fewer bottles will be needed. For our project, we would have used bigger bottles; however, very few people use such bottles, so we would have been waiting for a very long time. We made do with what was freely available. In some cases, you will find that certain 1-liter bottles work as a better form of insulation than the 1.5-liter bottles. This is due to the thickness of the bottle. You must assess your available materials and decide on your best option.

Tip 2: Spare bottles

It is quite a good idea to collect more bottles than you need. Some bottles in the completed greenhouse may get damaged over time, so you can easily replace them if there are extra bottles handy. This is the good thing about the design technique used for the greenhouse: each bottle strip is independent and can be replaced as needed. As long as there are plastic bottles, you will always have parts.

Tip 3: Ventilation

In our design, we have included a window. The reason for this is so that during those very hot days, we can open it so the inside will not get too hot and humid. It is always good for your plants to be aired. You should consider this at the design phase. However you are able to add a second or third window section if required. The technique used to construct the greenhouse gives you flexibility to make changes. For example we had installed a second window after a while as we felt that the greenhouse required more ventilation during the warmer seasons. The pictures below demonstrate before and after the second window section was installed.

Tip 4: Avoid erecting a greenhouse on soil ground
We have dedicated a section on building a base for the greenhouse. Any contact the timber frames have with soil will result in erosion over time. For this reason, you should create a base. It could be something similar to ours, or if you have something else in mind to prevent this contact with soil, give it a go.

Tip 5: Tar paper
For added protection, it would be ideal to put layers of tar paper on your base. This will prevent any escape of hot air, and it will protect the timber frames even more.

Tip 6: L-shaped supporting frames

These are a form of support to secure the greenhouse to the ground. It is highly possible that during windy periods, the greenhouse may move from its base. Adding these supporting frames will ensure that the greenhouse is stable. This is a matter of choice. The process is as follows:

a. Find L-shaped frames and pierce 3 holes on either side.
b. Dig a hole in the ground that is at least 20 cm and not much deeper.
c. Screw the frames to the four corners of the greenhouse.
d. Prepare some cement and then fill the holes up.
e. Even out the surface to appear smooth and leave to settle.

Tip 7: Added insulation and making it waterproof
You may find that having some gaps is unavoidable. Due to the shape of the bottles used to assemble the greenhouse, you will have some openings. To ensure that the heat doesn't escape from within the greenhouse, you may consider stapling a transparent nylon sheet over the internal walls. The picture below shows the application of the nylon on the major wall sections. This is recommended for those countries where it can get very cold.

The nylon sheets placed on the roof sections prevent rainwater from passing through. When placing the roof sections, make sure that the excess nylon on the ends is not tucked internally, but externally, so the rainwater can run off the roof.

Tip 8: Roof pitch

For countries where there is snowfall they must consider having a steeper vertical roof to allow for snow to slide off. It is recommended that you have a roof pitch (rise/run) of at least 12/9. It all depends on how much snow you have. In some cases you will have some rain after the snow which will increase the load on the roof, so the steeper you make it the better. See the below illustrations giving example of the roof pitch range.

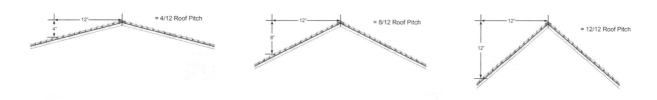

Tip 9: Greenhouse Location

As with any greenhouse you must ensure that it is exposed to as much sunlight as possible. It is especially important during installation to consider the sun's orientation during the winter, so that even low sun angles still benefit your greenhouse.

Chapter 8
A Peek inside our Greenhouse

How you decide to arrange the inside of your greenhouse and what you want to plant is entirely up to you. Make sure that you have enough space to plant your vegetables.

Ground Design

Ideally the greenhouse will be used to plant a variety of vegetables, so you will need to be able to move around easily. We have placed some slabs along the center of the greenhouse so that we have access to all parts of it. See the ground design and vegetables planted in our greenhouse below.

You can plant literally anything in your greenhouse, either plants or vegetables. Basically whatever you can grow in the warmer seasons, you can now grow in the colder seasons. In our greenhouse we have planted cucumbers, tomatoes, lettuce, eggplants, celery, basil, and chilies, which are the majority of the vegetables we eat.

Chapter 9
The Green Initiative

Reusing Bottle Caps

During the making of the plastic bottle greenhouse, you will have some waste such as cuttings from the bottles, bottle caps, and paper packaging around the bottles. Make sure that you properly dispose of these unless you can find an additional use for them. Bottle caps are used extensively for various crafts. With a little creative thinking, you will be surprised at what you can use them for.

A very good example of creativity is the renowned project by Mary Ellen Croteau, where she used over seven thousand plastic bottle caps to make a self-portrait. See below her plastic cap project. Mary has completed many interesting pieces of art, which can be seen on her site at www.maryellencroteau.net.

How To Build an Eco-Friendly Greenhouse

Overall we had managed to save approximately 2,500 bottle caps. We saved them, hoping that they would be of good use at some point. Then we posted a message on Facebook and luckily found someone who needed them.

At the time, a college, together with a local recycling company in Cyprus, in a joint effort to raise recycling awareness, assigned their students a task of collecting as many bottle caps as possible with the objective of exchanging them for a wheelchair. The college succeeded in collecting about sixty thousand caps, which is the equivalent of around 200 kg of material. This initial idea was first introduced in Turkey by Kushtrim Ahmeti of Kosovo in 2010, and it resulted in the distribution globally of nearly 1,500 wheelchairs.

What you think might be useless, others may find useful. A little research will help you unveil amazing solutions. Reuse is the most efficient and effective ecological approach through which we can help reduce our carbon footprint.

Promoting Green Initiative

Once your project is complete, why not communicate what you have done? Making your project known helps in promoting an eco-way of thinking and acting.

There are a number of persons and organizations willing to write about your project,— for example, local newspapers, magazines, and bloggers. Those companies that provided you with the support to help build your eco-project can be contacted, as they may promote it through their websites—especially recycling companies and plastic bottle companies.

APPENDIX A

Detailed Materials List

Timber:
200cm x4 pieces
350cm x9 pieces
195cm x4 pieces
250cm x8 pieces
143cm x8 pieces
77.5cm x2 pieces
340 cm x1 piece

Bamboo sticks
Roughly 215 canes

Saddle clips
Roughly 570

Short screws for saddle clips
Roughly 1140

Long screws for assembling sections
Roughly 60

Appendix B

Preparing the base area

This is a two-person activity, so you will need to find someone to help you.

1. Basically, you take a length of transparent hose (it must be at least the length of the surface area of your base from corner to corner diagonally plus a meter or so more), and you fill it up with water—not completely, but enough so that 85 to 90 percent of the hose has water. To ensure that there are no air bubbles, allow the water to come out through the other end of the hose when filling it; later you can spill some out to adjust the level.

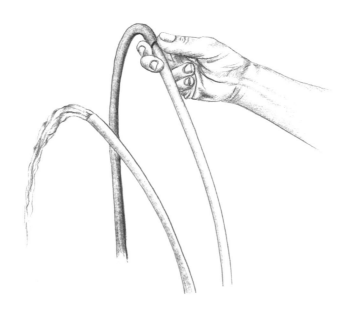

2. Place your thumbs or corks of some kind in the ends. Some people tie the two ends of the hose to two pieces of wood, holding them upright. This makes it easier to align the ends. In our case, we didn't do this since the heights involved were small.

3. When you place the two ends together, the water levels should be aligned to ensure they are level. Because the water level rises and falls as you change heights, make sure when moving the hose around that the water does not spill.

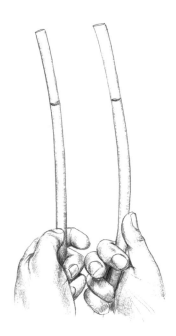

4. Once you have done this, you are then in a position to set your height levels. Four metal rods are pounded into the ground at the approximate corners of the base of the greenhouse. One person places a mark at a certain point on one of the four metal rods. This level is an approximation of the height of the base. It is used as the reference for the rods at the other three corners.

5. The water level is maintained at all times on the first mark. The first person holds one end of the hose at the mark on the first rod. The second person holds the other end of the hose and moves to each of the other rods, marking where the water level stops moving.

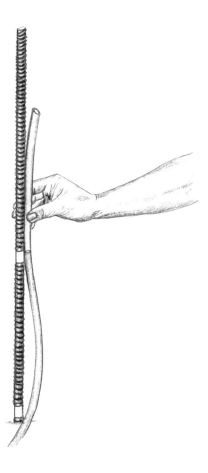

6. Once the four corners have been marked, then you can tie a string from each marked point to determine the height of your base. Make sure the string has been tied tightly to prevent it from moving. You may find that you need to adjust the height of the string, in which case, any movement up or down will also have to be applied to your other three points by the same distance.

7. Place a brick at each corner to approximate the height, and then determine if any adjustments are necessary. This will roughly determine how much you will be required to dig, bearing in mind the cement foundation thickness and perimeter.

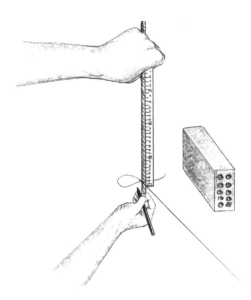

8. The string will act as a guide when making the base. The length and width of the base area depend on the dimensions of the greenhouse. This you can measure very easily, give or take one centimeter to be on the safe side.

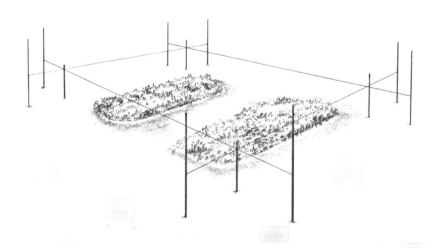

Measure the height and prepare the base area. See the image below.

Estimate your depth and dig a trench, taking into account the brick and cement layer. See the images below.

Mix the cement and start making the base, making sure the ceramic bricks are aligned to the height of the strings. See image below.

With regards to the cement follow the manufacturer's instructions on the bag.

About the Author

Andros Zacharia worked in the corporate sector for ten years until 2013, when he decided to take a leap in his career by embarking on new ventures independently. One of his endeavors involved setting up his own blog with a focus on eco-concepts and well-being. The eco-concept element looks at ways of improving and protecting our planet, and the well-being aspect is centered on encouraging people to avoid the systematic approach to life and adopt innovative methods that can help achieve a more content state of mind, achieving inner peace and outer purpose.

Andros lives in the village of Agia Anna in Cyprus which has a population of 150 people. The author was born in London but spent his formative years in Cyprus, where he returned after university in the United Kingdom. As part of his mission to expose himself to unique experiences, he spent some time living in Paris in order to learn a third language but also to experience a new culture. In addition to being multi-lingual, Andros does various forms of martial arts and intense fitness. He has enthusiasm for travelling and gourmet cuisine, as well as a peculiar interest in matchbox collecting.

Whilst working he was partly responsible for the implementation of paper recycling programmes as well as creating a more eco-friendly culture within his previous organization in Cyprus, which is one of the largest professional services companies in the world.

His guiding life motto has been to strive forward to be the change in society through a shift in his own thought process. He follows the philosophy that whatever seems impossible in life can become "I'm possible."

www.androszacharia.com

About the Book

Producing plastic is a cheap and straight forward process, however the disposal and decomposition of such materials is cumbersome and costly. Producing plastic is harming our environment primarily due the harmful elements that comprise it.

The concept of integrating plastic within a structure in a natural setting would appear to be incompatible. However this manual for making your own greenhouse offers a fresh perspective on re-using everyday materials such as the plastic bottle to be creative.

Our in-depth guide to making an eco-friendly greenhouse is not limited to its structure. The technique adopted can be applicable to anything else that you may be interested in constructing.

The book gives you a photographic and artistic account of how this was put together. There are useful tips and recommendations to enhance your eco-project as much as possible, making it both efficient and effective.

This guide to making an ecological greenhouse can be a source of inspiration for any DIY project, encouraging others to build and tell their stories.

Printed in the United States
By Bookmasters